higher-level thinking Questions
Life & Earth Sciences

questions by
Miguel Kagan
Christa Chapman

created and designed by
Miguel Kagan

illustrated by
Celso Rodriguez

© 1999 by *Kagan Publishing*

This book is published and distributed by *Kagan*. All rights are reserved by *Kagan*. No part of this publication may be reproduced or transmitted in any form by any means, electronic or mechanical, including photocopy, recording, or any information storage and retrieval system, without prior written permission from *Kagan*. The blackline masters included in this book may be duplicated only by classroom teachers who purchase the book, and only for use in their own classrooms. To obtain additional copies of this book, other *Kagan* publications, or information regarding *Kagan* professional development, contact *Kagan*.

Kagan Publishing
981 Calle Amanecer
San Clemente, CA 92673
1 (800) 933-2667
Fax: (949) 545-6301
www.KaganOnline.com

ISBN: 978-1-879097-51-3

Table of Contents

Introduction 2

1. Animals 27
2. Bugs 35
3. Dinosaurs 43
4. Endangered Species 51
5. Environmental Issues 59
6. Growing and Aging................. 67
7. Health and Nutrition 75
8. Human Body............. 83
9. Natural Disasters 91
10. Nature 99
11. Oceans 107
12. Plants and Trees................ 115
13. Rain Forest.............. 123
14. Seasons 131
15. Senses 139
16. Weather and Climate 147

1

> "I had six honest serving men
> They taught me all I knew:
> Their names were Where and What and When and Why and How and Who."
>
> — Rudyard Kipling

Introduction

In your hands you hold a powerful book. It is a member of a series of transformative blackline activity books. Between the covers, you will find questions, questions, and more questions! But these are no ordinary questions. These are the important kind—higher-level thinking questions—the kind that stretch your students' minds; the kind that release your students' natural curiosity about the world; the kind that rack your students' brains; the kind that instill in your students a sense of wonderment about your curriculum.

But we are getting a bit ahead of ourselves. Let's start from the beginning. Since this is a book of questions, it seems only appropriate for this introduction to pose a few questions—about the book and its underlying educational philosophy. So Mr. Kipling's Six Honest Serving Men, if you will, please lead the way:

What?
What are higher-level thinking questions?

This is a loaded question (as should be all good questions). Using our analytic thinking skills, let's break this question down into two smaller questions: 1) What is higher-level thinking? and 2) What are questions? When we understand the types of thinking skills and the types of questions, we can combine the best of both worlds, crafting beautiful questions to generate the range of higher-level thinking in our students!

Types of Thinking

There are many different types of thinking. Some types of thinking include:

- applying
- associating
- comparing
- contrasting
- defining
- elaborating
- empathizing
- experimenting
- generalizing
- investigating
- making analogies
- planning
- prioritizing
- recalling
- reflecting
- reversing
- sequencing
- summarizing
- synthesizing
- assessing
- augmenting
- connecting
- decision-making
- drawing conclusions
- eliminating
- evaluating
- explaining
- inferring consequences
- inventing
- memorizing
- predicting
- problem-solving
- reducing
- relating
- role-taking
- substituting
- symbolizing
- understanding
- thinking about thinking (metacognition)

This is quite a formidable list. It's nowhere near complete. Thinking is a big, multifaceted phenomenon. Perhaps the most widely recognized system for classifying thinking and classroom questions is Benjamin Bloom's Taxonomy of Thinking Skills. Bloom's Taxonomy classifies thinking skills into six hierarchical levels. It begins with the lower levels of thinking skills and moves up to higher-level thinking skills: 1) Knowledge, 2) Comprehension, 3) Application, 4) Analysis, 5) Synthesis, 6) Evaluation. See Bloom's Taxonomy on the following page.

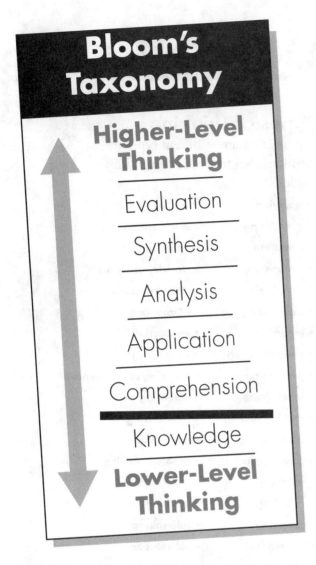

In education, the term "higher-level thinking" often refers to the higher levels of Mr. Bloom's taxonomy. But Bloom's Taxonomy is but one way of organizing and conceptualizing the various types of thinking skills.

There are many ways we can cut the thinking skills pie. We can alternatively view the many different types of thinking skills as, well…many different skills. Some thinking skills may be hierarchical. Some may be interrelated. And some may be relatively independent.

In this book, we take a pragmatic, functional approach. Each type of thinking skill serves a different function. So called "lower-level" thinking skills are very useful for certain purposes. Memorizing and understanding information are invaluable skills that our students will use throughout their lives. But so too are many of the "higher-level" thinking skills on our list. The more facets of students' thinking skills we develop, the better we prepare them for lifelong success.

Because so much classroom learning heretofore has focused on the "lower rungs" of the thinking skills ladder—knowledge and comprehension, or memorization and understanding—in this series of books we have chosen to focus on questions to generate "higher-level" thinking. This book is an attempt to correct the imbalance in the types of thinking skills developed by classroom questions.

Types of Questions

As we ask questions of our students, we further promote cognitive development when we use Fat questions, Low-Consensus questions, and True questions.

Fat Questions vs. Skinny Questions

Skinny questions are questions that require a skinny answer. For example, after reading a poem, we can ask: "Did you like the poem?" Even though this question could be categorized as an Evaluation question—Bloom's highest level of thinking— it can be answered with one monosyllabic word: "Yes" or "No." How much thinking are we actually generating in our students?

We can reframe this question to make it a fat question: "What things did you like about the poem? What things did you dislike?" Notice no short answer will do. Answering this fattened-up question requires more elaboration. These fat questions presuppose not that there is only one thing but things plural that the student liked and things that she did not like. Making things plural is one way to make skinny questions fat. Students stretch their minds to come up with multiple ideas or solutions. Other easy ways to make questions fat is to add "Why or why not?" or "Explain" or "Describe" or "Defend your position" to the end of a question. These additions promote elaboration beyond a skinny answer. Because language and thought are intimately intertwined, questions that require elaborate responses stretch students' thinking: They grapple to articulate their thoughts.

The type of questions we ask impact not just the type of thinking we develop in our students, but also the depth of thought. Fat questions elicit fat responses. Fat responses develop both depth of thinking and range of thinking skills. The questions in this book are designed to elicit fat responses—deep and varied thinking.

High-Consensus Questions vs. Low-Consensus Questions

A high-consensus question is one to which most people would give the same response, usually a right or wrong answer. After learning about sound, we can ask our students: "What is the name of a room specially designed to improve acoustics for the audience?" This is a high-consensus question. The answer (auditorium) is either correct or incorrect.

Compare the previous question with a low-consensus question: "If you were going to build an auditorium, what special design features would you take into consideration?" Notice, to the low-consensus question there is no right or wrong answer. Each person formulates his or her unique response. To answer, students must apply what they learned, use their ingenuity and creativity.

High-consensus questions promote convergent thinking. With high-consensus questions we strive to direct students *what to think*. Low-consensus questions promote divergent thinking, both critical and creative. With low-consensus questions we strive to develop students' *ability to think*. The questions in this book are low-consensus questions designed to promote independent, critical and creative thought.

True Questions vs. Review Questions

We all know what review questions are. They're the ones in the back of every chapter and unit. Review questions ask students to regurgitate previously stated or learned information. For example, after learning about the rain forest we may ask: "What percent of the world's oxygen does the rain forest produce?" Students can go back a few pages in their books or into their memory banks and pull out the answer. This is great if we are working on memorization skills, but does little to develop "higher-order" thinking skills.

True questions, on the other hand, are meaningful questions—questions to which we do not know the answer. For example: "What might happen if all the world's rain forests were cut down?" This is a hypothetical; we don't know the answer but considering the question forces us to think. We infer some logical consequences based on what we know. The goal of true questions is not a correct answer, but the thinking journey students take to create a meaningful response. True questions are more representative of real life. Seldom is there a black and white answer. In life, we struggle with ambiguity, confounding variables, and uncertain outcomes. There are millions of shades of gray. True questions prepare students to deal with life's uncertainties.

When we ask a review question, we know the answer and are checking to see if the student does also. When we ask a true question, it is truly a question. We don't necessarily know the answer and neither does the student. True questions are

> **Education is not the filling of a pail, but the lighting of a fire.**
> — William Butler Yeats

Types of Questions

Skinny → **Fat**
- Short Answer
- Shallow Thinking

- Elaborated Answer
- Deep Thinking

High-Consensus → **Low-Consensus**
- Right or Wrong Answer
- Develops Convergent Thinking
- "What" to Think

- No Single Correct Answer
- Develops Divergent Thinking
- "How" to Think

Review → **True**
- Asker Knows Answer
- Checking for Correctness

- Asker Doesn't Know Answer
- Invitation to Think

often an invitation to think, ponder, speculate, and engage in a questioning process.

We can use true questions in the classroom to make our curriculum more personally meaningful, to promote investigation, and awaken students' sense of awe and wonderment in what we teach. Many questions you will find in this book are true questions designed to make the content provocative, intriguing, and personally relevant.

The box above summarizes the different types of questions. The questions you will find in this book are a move away from skinny, high-consensus, review questions toward fat, low-consensus true questions. As we ask these types of questions in our class, we transform even mundane content into a springboard for higher-level thinking. As we integrate these question gems into our daily lessons, we create powerful learning experiences. ***We do not fill our students' pails with knowledge; we kindle their fires to become lifetime thinkers.***

Why?
Why should I use higher-level thinking questions in my classroom?

As we enter the new millennium, major shifts in our economic structure are changing the ways we work and live. The direction is increasingly toward an information-based, high-tech economy. The sum of our technological information is exploding. We could give you a figure how rapidly information is doubling, but by the time you read this, the number would be outdated! No kidding.

But this is no surprise. This is our daily reality. We see it around us everyday and on the news: cloning, gene manipulation, e-mail, the Internet, Mars rovers, electric cars, hybrids, laser surgery, CD-ROMs, DVDs. All around us we see the wheels of progress turning: New discoveries, new technologies, a new societal knowledge and information base. New jobs are being created today in fields that simply didn't exist yesterday.

How do we best prepare our students for this uncertain future—a future in which the only constant will be change? As we are propelled into a world of ever-increasing change, what is the relative value of teaching students facts versus thinking skills? This point becomes even more salient when we realize that students cannot master everything, and many facts will soon become obsolete. Facts become outdated or irrelevant. Thinking skills are for a lifetime. Increasingly, how we define educational success will be away from the quantity of information mastered. Instead, we will define success as our students' ability to generate questions, apply, synthesize, predict, evaluate, compare, categorize.

If we as a professionals are to proactively respond to these societal shifts, thinking skills will become central to our curriculum. Whether we teach thinking skills directly, or we integrate them into our curriculum, the power to think is the greatest gift we can give our students!

We believe the questions you will find in this book are a step in the direction of preparing students for lifelong success. The goal is to develop independent thinkers who are critical and creative, regardless of the content. We hope the books in this series are more than sets of questions. We provide them as a model approach to questioning in the classroom.

On pages 8 and 9, you will find Questions to Engage Students' Thinking Skills. These pages contain numerous types of thinking and questions designed to engage each thinking skill. As you make your own questions for your students with your own content, use these question starters to help you frame

> **Virtually the only predictable trend is continuing change.**
> — Dr. Linda Tsantis, Creating the Future

your questions to stimulate various facets of your students' thinking skills. Also let your students use these question starters to generate their own higher-level thinking questions about the curriculum.

Who?
Who is this book for?

This book is for you and your students, but mostly for your students. It is designed to help make your job easier. Inside you will find hundreds of ready-to-use reproducible questions. Sometimes in the press for time we opt for what is easy over what is best. These books attempt to make easy what is best. In this treasure chest, you will find hours and hours of timesaving ready-made questions and activities.

Place Higher-Level Thinking In Your Students' Hands

As previously mentioned, this book is even more for your students than for you. As teachers, we ask a tremendous number of questions. Primary teachers ask 3.5 to 6.5 questions per minute! Elementary teachers average 348 questions a day. How many questions would you predict our students ask? Researchers asked this question. What they found was shocking: Typical students ask approximately one question per month.* One question per month!

Although this study may not be representative of your classroom, it does suggest that in general, as teachers we are missing out on a very powerful force—student-generated questions. The capacity to answer higher-level thinking questions is

---

* Myra & David Sadker, "Questioning Skills" in *Classroom Teaching Skills*, 2nd ed. Lexington, MA: D.C. Heath & Co., 1982.

Questions to Engage Students' Thinking Skills

Analyzing
- How could you break down…?
- What components…?
- What qualities/characteristics…?

Applying
- How is _____ an example of…?
- What practical applications…?
- What examples…?
- How could you use…?
- How does this apply to…?
- In your life, how would you apply…?

Assessing
- By what criteria would you assess…?
- What grade would you give…?
- How could you improve…?

Augmenting/Elaborating
- What ideas might you add to…?
- What more can you say about…?

Categorizing/Classifying/Organizing
- How might you classify…?
- If you were going to categorize…?

Comparing/Contrasting
- How would you compare…?
- What similarities…?
- What are the differences between…?
- How is _____ different…?

Connecting/Associating
- What do you already know about…?
- What connections can you make between…?
- What things do you think of when you think of…?

Decision-Making
- How would you decide…?
- If you had to choose between…?

Defining
- How would you define…?
- In your own words, what is…?

Describing/Summarizing
- How could you describe/summarize…?
- If you were a reporter, how would you describe…?

Determining Cause/Effect
- What is the cause of…?
- How does _____ effect _____?
- What impact might…?

Drawing Conclusions/Inferring Consequences
- What conclusions can you draw from…?
- What would happen if…?
- What would have happened if…?
- If you changed _____, what might happen?

Eliminating
- What part of _____ might you eliminate?
- How could you get rid of…?

Evaluating
- What is your opinion about…?
- Do you prefer…?
- Would you rather…?
- What is your favorite…?
- Do you agree or disagree…?
- What are the positive and negative aspects of…?
- What are the advantages and disadvantages…?
- If you were a judge…?
- On a scale of 1 to 10, how would you rate…?
- What is the most important…?
- Is it better or worse…?

Explaining
- How can you explain…?
- What factors might explain…?

Experimenting
- How could you test…?
- What experiment could you do to…?

Generalizing
- What general rule can…?
- What principle could you apply…?
- What can you say about all…?

Interpreting
- Why is ____ important?
- What is the significance of…?
- What role…?
- What is the moral of…?

Inventing
- What could you invent to…?
- What machine could…?

Investigating
- How could you find out more about…?
- If you wanted to know about…?

Making Analogies
- How is ____ like ____?
- What analogy can you invent for…?

Observing
- What observations did you make about…?
- What changes…?

Patterning
- What patterns can you find…?
- How would you describe the organization of…?

Planning
- What preparations would you…?

Predicting/Hypothesizing
- What would you predict…?
- What is your theory about…?
- If you were going to guess…?

Prioritizing
- What is more important…?
- How might you prioritize…?

Problem-Solving
- How would you approach the problem?
- What are some possible solutions to…?

Reducing/Simplifying
- In a word, how would you describe…?
- How can you simplify…?

Reflecting/Metacognition
- What would you think if…?
- How can you describe what you were thinking when…?

Relating
- How is ____ related to ____?
- What is the relationship between…?
- How does ____ depend on ____?

Reversing/Inversing
- What is the opposite of…?

Role-Taking/Empathizing
- If you were (someone/something else)…?
- How would you feel if…?

Sequencing
- How could you sequence…?
- What steps are involved in…?

Substituting
- What could have been used instead of…?
- What else could you use for…?
- What might you substitute for…?
- What is another way…?

Symbolizing
- How could you draw…?
- What symbol best represents…?

Synthesizing
- How could you combine…?
- What could you put together…?

a wonderful skill we can give our students, as is the skill to solve problems. Arguably more important skills are the ability to find problems to solve and formulate questions to answer. If we look at the great thinkers of the world—the Einsteins, the Edisons, the Freuds—their thinking is marked by a yearning to solve tremendous questions and problems. It is this questioning process that distinguishes those who illuminate and create our world from those who merely accept it.

Make Learning an Interactive Process

Higher-level thinking is not just something that occurs between students' ears! Students benefit from an interactive process. This basic premise underlies the majority of activities you will find in this book.

As students discuss questions and listen to others, they are confronted with differing perspectives and are pushed to articulate their own thinking well beyond the level they could attain on their own. Students too have an enormous capacity to mediate each other's learning. When we heterogeneously group students to work together, we create an environment to move students through their zone of proximal development. We also provide opportunities for tutoring and leadership. Verbal interaction with peers in cooperative groups adds a dimension to questions not available with whole-class questions and answers.

> **Asking a good question requires students to think harder than giving a good answer.**
> — Robert Fisher, Teaching Children to Learn

Reflect on this analogy: If we wanted to teach our students to catch and throw, we could bring in one tennis ball and take turns throwing it to each student and having them throw it back to us. Alternatively, we could bring in twenty balls and have our students form small groups and have them toss the ball back and forth to each other. Picture the two classrooms: One with twenty balls being caught at any one moment, and the other with just one. In which class would students better and more quickly learn to catch and throw?

The same is true with thinking skills. When we make our students more active participants in the learning process, they are given dramatically more opportunities to produce their own thought and to strengthen their own thinking skills. Would you rather have one question being asked and answered at any one moment in your class, or twenty? Small groups mean more questioning and more thinking. Instead of rarely answering a teacher question or rarely generating their own question, asking and answering questions becomes a regular part of your students' day. It is through cooperative interaction that we truly turn our classroom into a higher-level think tank. The associated personal and social benefits are invaluable.

When?
When do I use higher-level thinking questions?

Do I use these questions at the beginning of the lesson, during the lesson, or after? The answer, of course, is all of the above.

Use these questions or your own thinking questions at the beginning of the lesson to provide a motivational set for the lesson. Pique students' interest about the content with some provocative questions: "What would happen if we didn't have gravity?" "Why did Pilgrims get along with some Native Americans, but not others?" "What do you think this book will be about?" Make the content personally relevant by bringing in students' own knowledge, experiences, and feelings about the content: "What do you know about spiders?" "What things do you like about mystery stories?" "How would you feel if explorers invaded your land and killed your family?" "What do you wonder about electricity?"

Use the higher-level thinking questions throughout your lessons. Use the many questions and activities in this book not as a replacement of your curriculum, but as an additional avenue to explore the content and stretch students' thinking skills.

Use the questions after your lesson. Use the higher-level thinking questions, a journal writing activity, or the question starters as an extension activity to your lesson or unit.

Or just use the questions as stand-alone sponge activities for students or teams who have finished their work and need a challenging project to work on.

It doesn't matter when you use them, just use them frequently. As questioning becomes a habitual part of the classroom day, students' fear of asking silly questions is diminished. As the ancient Chinese proverb states, "Those who ask a silly question may seem a fool for five minutes, but those who do not ask remain a fool for life."

> **The important thing is to never stop questioning.**
> — Albert Einstein

As teachers, we should make a conscious effort to ensure that a portion of the many questions we ask on a daily basis are those that move our students beyond rote memorization. When we integrate higher-level thinking questions into our daily lessons, we transform our role from transmitters of knowledge to engineers of learning.

Where?
Where should I keep this book?

Keep it close by. Inside there are 16 sets of questions. Pull it out any time you teach these topics or need a quick, easy, fun activity or journal writing topic.

How?
How do I get the most out of this book?

In this book you will find 16 topics arranged alphabetically. For each topic there are reproducible pages for: 1) 16 Question Cards, 2) a Journal Writing activity page, 3) and a Question Starters activity page.

1. Question Cards

The Question Cards are truly the heart of this book. There are numerous ways the Question Cards can be used. After the other activity pages are introduced, you will find a description of a variety of engaging formats to use the Question Cards.

Specific and General Questions

Some of the questions provided in this book series are content-specific and others are content-free. For example, the literature questions in the Literature books are content-specific. Questions for the Great Kapok Tree deal specifically with that literature selection. Some language arts questions in the Language Arts book, on the other hand, are content-free. They are general questions that can be used over and over again with new content. For example, the Book Review questions can be used after reading any book. The Story Structure questions can be used after reading any story. You can tell by glancing at the title of the set and some of the questions whether the set is content-specific or content-free.

A Little Disclaimer

Not all of the "questions" on the Question Cards are actually questions. Some instruct students to do something. For example, "Compare and contrast…" We can also use these directives to develop the various facets of students' thinking skills.

The Power of Think Time

As you and your students use these questions, don't forget about the power of Think Time! There are two different think times. The first is the time between the question and the response. The second is the time between the response and feedback on the response. Think time has been shown to greatly enhance the quality of student thinking. If students are not pausing for either think time, or doing it too briefly, emphasize its importance. Five little seconds of silent think time after the question and five more seconds before feedback are proven, powerful ways to promote higher-level thinking in your class.

Use Your Question Cards for Years

For attractive Question Cards that will last for years, photocopy them on color card-stock paper and laminate them. To save time, have the Materials Monitor from each team pick up one card set, a pair of scissors for the team, and an envelope or rubber band. Each team cuts out their own set of Question Cards. When they are done with the activity, students can place the Question Cards in the envelope and write the name of the set on the envelope or wrap the cards with a rubber band for storage.

2. Journal Question

The Journal Writing page contains one of the 16 questions as a journal writing prompt. You can substitute any question, or use one of your own. The power of journal writing cannot be overstated. The act of writing takes longer than speaking and thinking. It allows the brain time to make deep connections to the content. Writing requires the writer to present his or her response in a clear, concise language. Writing develops both strong thinking and communication skills.

A helpful activity before journal writing is to have students discuss the question in pairs or in small teams. Students discuss their ideas and what they plan to write. This little prewriting activity ignites ideas for those students who stare blankly at their Journal Writing page. The interpersonal interaction further helps students articulate what they are thinking about the topic and invites students to delve deeper into the topic.

Tell students before they write that they will share their journal entries with a partner or with their team. This motivates many students to improve their entry. Sharing written responses also promotes flexible thinking with open-ended questions, and allows students to hear their peers' responses, ideas and writing styles.

Have students keep a collection of their journal entries in a three-ring binder. This way you can collect them if you wish for assessment or have students go back to reflect on their own learning. If you are using questions across the curriculum, each subject can have its own journal or own section within the binder. Use the provided blackline on the following page for a cover for students' journals or have students design their own.

3. Question Starters

The Question Starters activity page is designed to put the questions in the hands of your students. Use these question starters to scaffold your students' ability to write their own thinking questions. This page includes eight question starters to direct students to generate questions across the levels and types of thinking. This Question Starters activity page can be used in a few different ways:

Individual Questions

Have students independently come up with their own questions. When done, they can trade their questions with a partner. On a separate sheet of paper students answer their partners' questions. After answering, partners can share how they answered each other's questions.

JOURNAL

My Best Thinking

This Journal Belongs to

Pair Questions
Students work in pairs to generate questions to send to another pair. Partners take turns writing each question and also take turns recording each answer. After answering, pairs pair up to share how they answered each other's questions.

Team Questions
Students work in teams to generate questions to send to another team. Teammates take turns writing each question and recording each answer. After answering, teams pair up to share how they answered each other's questions.

Teacher-Led Questions
For young students, lead the whole class in coming up with good higher-level thinking questions.

Teach Your Students About Thinking and Questions
An effective tool to improve students' thinking skills is to teach students about the types of thinking skills and types of questions. Teaching students about the types of thinking skills improves their metacognitive abilities. When students are aware of the types of thinking, they may more effectively plan, monitor, and evaluate their own thinking. When students understand the types of questions and the basics of question construction, they are more likely to create effective higher-level thinking questions. In doing so they develop their own thinking skills and the thinking of classmates as they work to answer each other's questions.

Table of Activities

The Question Cards can be used in a variety of game-like formats to forge students' thinking skills. They can be used for cooperative team and pair work, for whole-class questioning, for independent activities, or at learning centers. On the following pages you will find numerous excellent options to use your Question Cards. As you use the Question Cards in this book, try the different activities listed below to add novelty and variety to the higher-level thinking process.

Team Activities
1. Question Commander 16
2. Fan-N-Pick 18
3. Spin-N-Think 18
4. Three-Step Interview 19
5. Team Discussion 19
6. Think-Pair-Square 20
7. Question-Write-RoundRobin 20

Class Activities
1. Mix-Pair-Discuss 21
2. Think-Pair-Share 21
3. Inside-Outside Circle 22
4. Question & Answer 22
5. Numbered Heads Together 23

Pair Activities
1. RallyRobin 23
2. Pair Discussion 24
3. Question-Write-Share-Discuss 24

Individual Activities
1. Journal Writing 25
2. Independent Answers 25

Learning Centers
1. Question Card Center 26
2. Journal Writing Center 26
3. Question Starters Center 26

Higher-Level Thinking Question Card
Activities

team activity #1

Question Commander

Preferably in teams of four, students shuffle their Question Cards and place them in a stack, questions facing down, so that all teammates can easily reach the Question Cards. Give each team a Question Commander set of instructions (blackline provided on following page) to lead them through each question.

Student One becomes the Question Commander for the first question. The Question Commander reads the question aloud to the team, then asks the teammates to think about the question and how they would answer it. After the think time, the Question Commander selects a teammate to answer the question. The Question Commander can spin a spinner or roll a die to select who will answer. After the teammate gives the answer, Question Commander again calls for think time, this time asking the team to think about the answer. After the think time, the Question Commander leads a team discussion in which any teammember can contribute his or her thoughts or ideas to the question, or give praise or reactions to the answer.

When the discussion is over, Student Two becomes the Question Commander for the next question.

Question Commander
Instruction Cards

Question Commander

1. **Ask the Question:** Question Commander reads the question to the team.
2. **Think Time:** "Think of your best answer."
3. **Answer the Question:** The Question Commander selects a teammate to answer the question.
4. **Think Time:** "Think about how you would answer differently or add to the answer."
5. **Team Discussion:** As a team, discuss other possible answers or reactions to the answer given.

Question Commander

1. **Ask the Question:** Question Commander reads the question to the team.
2. **Think Time:** "Think of your best answer."
3. **Answer the Question:** The Question Commander selects a teammate to answer the question.
4. **Think Time:** "Think about how you would answer differently or add to the answer."
5. **Team Discussion:** As a team, discuss other possible answers or reactions to the answer given.

Question Commander

1. **Ask the Question:** Question Commander reads the question to the team.
2. **Think Time:** "Think of your best answer."
3. **Answer the Question:** The Question Commander selects a teammate to answer the question.
4. **Think Time:** "Think about how you would answer differently or add to the answer."
5. **Team Discussion:** As a team, discuss other possible answers or reactions to the answer given.

Question Commander

1. **Ask the Question:** Question Commander reads the question to the team.
2. **Think Time:** "Think of your best answer."
3. **Answer the Question:** The Question Commander selects a teammate to answer the question.
4. **Think Time:** "Think about how you would answer differently or add to the answer."
5. **Team Discussion:** As a team, discuss other possible answers or reactions to the answer given.

team activity #2

Fan-N-Pick

In a team of four, Student One fans out the question cards, and says, "Pick a card, any card!" Student Two picks a card and reads the question out loud to teammates. After five seconds of think time, Student Three gives his or her answer. After another five seconds of think time, Student Four paraphrases, praises, or adds to the answer given. Students rotate roles for each new round.

team activity #3

Spin-N-Think

Spin-N-Think spinners are available from Kagan to lead teams through the steps of higher-level thinking. Students spin the Spin-N-Think™ spinner to select a student at each stage of the questioning to: 1) ask the question, 2) answer the question, 3) paraphrase and praise the answer, 4) augment the answer, and 5) discuss the question or answer. The Spin-N-Think™ game makes higher-level thinking more fun, and holds students accountable because they are often called upon, but never know when their number will come up.

team activity #4

Three-Step Interview

After the question is read to the team, students pair up. The first step is an interview in which one student interviews the other about the question. In the second step, students remain with their partner but switch roles: The interviewer becomes the interviewee. In the third step, the pairs come back together and each student in turn presents to the team what their partner shared. Three-Step Interview is strong for individual accountability, active listening, and paraphrasing skills.

team activity #5

Team Discussion

Team Discussion is an easy and informal way of processing the questions: Students read a question and then throw it open for discussion. Team Discussion, however, does not ensure that there is individual accountability or equal participation.

team activity #6

Think-Pair-Square

One student reads a question out loud to teammates. Partners on the same side of the table then pair up to discuss the question and their answers. Then, all four students come together for an open discussion about the question.

team activity #7

Question-Write-RoundRobin

Students take turns asking the team the question. After each question is asked, each student writes his or her ideas on a piece of paper. After students have finished writing, in turn they share their ideas. This format creates strong individual accountability because each student is expected to develop and share an answer for every question.

Higher-Level Thinking Questions for Life and Earth Sciences
Kagan Publishing • 1 (800) 933-2667 • www.KaganOnline.com

class activity #1

Mix-Pair-Discuss

Each student gets a different Question Card. For 16 to 32 students, use two sets of questions. In this case, some students may have the same question which is OK. Students get out of their seats and mix around the classroom. They pair up with a partner. One partner reads his or her Question Card and the other answers. Then they switch roles. When done they trade cards and find a new partner. The process is repeated for a predetermined amount of time. The rule is students cannot pair up with the same partner twice. Students may get the same questions twice or more, but each time it is with a new partner. This strategy is a fun, energizing way to ask and answer questions.

class activity #2

Think-Pair-Share

Think-Pair-Share is teacher-directed. The teacher asks the question, then gives students think time. Students then pair up to share their thoughts about the question. After the pair discussion, one student is called on to share with the class what was shared in his or her pair. Think-Pair-Share does not provide as much active participation for students as Think-Pair-Square because only one student is called upon at a time, but is a nice way to do whole-class sharing.

class activity #3

Inside-Outside Circle

Each student gets a Question Card. Half of the students form a circle facing out. The other half forms a circle around the inside circle; each student in the outside circle faces one student in the inside circle. Students in the outside circle ask inside circle students a question. After the inside circle students answer the question, students switch roles questioning and answering. After both have asked and answered a question, they each praise the other's answers and then hold up a hand indicating they are finished. When most students have a hand up, have students trade cards with their partner and rotate to a new partner. To rotate, tell the outside circle to move to the left. This format is a lively and enjoyable way to ask questions and have students listen to the thinking of many classmates.

class activity #4

Question & Answer

This might sound familiar: Instead of giving students the Question Cards, the teacher asks the questions and calls on one student at a time to answer. This traditional format eliminates simultaneous, cooperative interaction, but may be good for introducing younger students to higher-level questions.

class activity #5

Numbered Heads Together

Students number off in their teams so that every student has a number. The teacher asks a question. Students put their "heads together" to discuss the question. The teacher then calls on a number and selects a student with that number to share what his or her team discussed.

pair activity #1

RallyRobin

Each pair gets a set of Question Cards. Student A in the pair reads the question out loud to his or her partner. Student B answers. Partners take turns asking and answering each question.

pair activity #2

Pair Discussion

Partners take turns asking the question. The pair then discusses the answer together. Unlike RallyRobin, students discuss the answer. Both students contribute to answering and to discussing each other's ideas.

pair activity #3

Question-Write-Share-Discuss

One partner reads the Question Card out loud to his or her teammate. Both students write down their ideas. Partners take turns sharing what they wrote. Partners discuss how their ideas are similar and different.

individual activity #1

Journal Writing

Students pick one Question Card and make a journal entry or use the question as the prompt for an essay or creative writing. Have students share their writing with a partner or in turn with teammates.

individual activity #2

Independent Answers

Students each get their own set of Questions Cards. Pairs or teams can share a set of questions, or the questions can be written on the board or put on the overhead projector. Students work by themselves to answer the questions on a separate sheet of paper. When done, students can compare their answers with a partner, teammates, or the whole class.

Center Ideas

1. Question Card Center
At one center, have the Question Cards and a Spin-N-Think™ spinner, Question Commander instruction card, or Fan-N-Pick instructions. Students lead themselves through the thinking questions. For individual accountability, have each student record their own answer for each question.

2. Journal Writing Center
At a second center, have a Journal Writing activity page for each student. Students can discuss the question with others at their center, then write their own journal entry. After everyone is done writing, students share what they wrote with other students at their center.

3. Question Starters Center
At a third center, have a Question Starters page. Split the students at the center into two groups. Have both groups create thinking questions using the Question Starters activity page. When the groups are done writing their questions, they trade questions with the other group at their center. When done answering each other's questions, two groups pair up to compare their answers.

Animals
higher-level thinking questions

> **Man's greatness lies in his power of thought.**
>
> — Blaise Pascal

Animals
Question Cards

Animals

1 Humans are animals too, just with more sophisticated brains. Do you agree or disagree with this statement?

Animals

2 How are dogs and cats similar? How are they different? Compare and contrast dogs and cats.

Animals

3 Should animals be kept in zoos for people's entertainment and education or is it cruelty to animals? Describe your position.

Animals

4 If you could speak to any animal, which would it be? What would you ask him or her?

Animals
Question Cards

Animals

5 What animal are you most afraid of? Why?

Animals

6 Do animals have feelings? Explain your answer.

Animals

7 Which animal do you think is the smartest? Why?

Animals

8 Why do you think people keep animals of all types as their pets? Do you have any pets? Why or why not?

Animals
Question Cards

Animals

9 What do animals teach us?

Animals

10 Are you opposed to fur coats and leather boots, jackets, belts and purses? Describe your position.

Animals

11 Many dogs and cats end up in the pound and are put to sleep. If you were the head of the Humane Society, what would you do?

Animals

12 There are animals of all sorts: big, small, short, tall, furry, scaly, flying, swimming, colorful, plain. Why are there so many different kinds of animals?

Animals
Question Cards

Animals

13 If a species is endangered, should people intervene to prevent extinction? If not, why not? If so, what should we do?

Animals

14 Should cosmetic testing and medical research be done on animals? Why or why not?

Animals

15 Would you rather be a bird, fish, mammal, reptile, or amphibian? Why?

Animals

16 Do animals talk to each other? Explain.

Higher-Level Thinking Questions for Life and Earth Sciences
Kagan Publishing • 1 (800) 933-2667 • www.KaganOnline.com

Animals
Journal Writing Question

Write your response to the question below.
Be ready to share your response.

Should cosmetic testing and medical research be done on animals? Why or why not?

Animals
Question Starters

Use the question starters below to create complete questions.
Send your questions to a partner or to another team to answer.

1. What plans would you make if

2. How could you test

3. If you were a dog

4. What is the most important

5. Would you rather

6. What is an example of

7. How are animals

8. Where might

Higher-Level Thinking Questions for Life and Earth Sciences
Kagan Publishing • 1 (800) 933-2667 • www.KaganOnline.com

Bugs
higher-level thinking questions

When thoughts arise,
then do all things arise.
When thoughts vanish,
then do all things vanish.

— Huang Po

Bugs
Question Cards

Bugs

1 How would your life be different if you were the size of a bug?

Bugs

2 Are insects loners or social creatures? Explain your answer.

Bugs

3 People swat flies, squish ants, and exterminate termites. Do we have the right? Why or why not?

Bugs

4 If you could be one insect for a week, which insect would you be? Why?

Bugs
Question Cards

Bugs

5. Do you think insects have feelings and thoughts or do they act only on instinct? Explain.

Bugs

6. Caterpillars go through a metamorphosis to become beautiful butterflies. Do humans go through a metamorphosis?

Bugs

7. "Busy as a bee," and "social butterfly" are common bug metaphors. What do they mean? Make up a new bug metaphor and explain what it means.

Bugs

8. Insects have three pairs of legs. What would be the positives and negatives of having six legs?

Higher-Level Thinking Questions for Life and Earth Sciences

Bugs
Question Cards

Bugs

9 There are about one million known species of insects. Millions more are probably not yet identified. Why do you think there are so many different types of insects? Why have insects done so well on Earth?

Bugs

10 How would the world be different if there were no bugs? Would it be better or worse? Explain.

Bugs

11 How are insects similar to other animals? How are they different?

Bugs

12 What would happen if bugs were as big as dogs?

Bugs
Question Cards

Bugs

13 The word "bug" also means to annoy or bother. The word "pest" means nuisance. What does this say about how people feel about insects? Do bugs bug you?

Bugs

14 Many insects can lift or drag 20 times their own weight. What would you be able to lift or drag? What would you do if you were that strong?

Bugs

15 Spiderman had the powers of a spider. What insect would you like the powers of? Describe how you'd use your powers.

Bugs

16 List different kinds of bugs. If you could organize the bugs into categories, which categories would you use?

40

Higher-Level Thinking Questions for Life and Earth Sciences
Kagan Publishing • 1 (800) 933-2667 • www.KaganOnline.com

Bugs

Journal Writing Question

Write your response to the question below.
Be ready to share your response.

Spiderman had the powers of a spider. What insect would you like the powers of? Describe how you'd use your powers.

Bugs
Question Starters

Use the question starters below to create complete questions. Send your questions to a partner or to another team to answer.

1. If you were a bug

2. Why do bugs

3. In your life

4. What is the relationship

5. How could you represent

6. What do you know about

7. What would be worse

8. Why do you think

Dinosaurs
higher-level thinking questions

"Thoughts have power; thoughts are energy. And you can make your world or break it by your own thinking."

— Susan Taylor

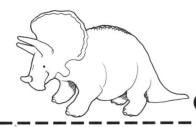

Dinosaurs
Question Cards

Dinosaurs

1 There are lots of theories why dinosaurs became extinct. Some say it got too cold and the cold-blooded dinosaurs died. Some think an asteroid hit the earth. Some think a star exploded and they were killed by cosmic rays. Why do you think dinosaurs became extinct?

Dinosaurs

2 If we could resurrect dinosaurs using their DNA like in the movie "Jurassic Park," should we? Why or why not?

Dinosaurs

3 How would your life be different if you lived 100 million years ago, in the time of dinosaurs?

Dinosaurs

4 If you could be any dinosaur for a day, which dinosaur would you be? Why?

Higher-Level Thinking Questions for Life and Earth Sciences
Kagan Publishing • 1 (800) 933-2667 • www.KaganOnline.com

Dinosaurs
Question Cards

Dinosaurs

5 If you wanted to make a movie about dinosaurs, how would you make them seem real in the movie?

Dinosaurs

6 Which living animal reminds you most of the dinosaur? How are they similar? How are they different?

Dinosaurs

7 If dinosaurs did not become extinct, how would the world be different today?

Dinosaurs

8 How do humans know so much about dinosaurs when we didn't exist when they were around?

Dinosaurs
Question Cards

Dinosaurs

9 Dinosaurs are like _____ because...

Dinosaurs

10 Dinosaurs once ruled the earth and now are extinct. It can be said humans now rule the earth. Do you think humans will become extinct too? Why or why not?

Dinosaurs

11 The brachiosaurus got over 70 feet long and weighed as much as 70 tons. What would it be like to be that big?

Dinosaurs

12 Fossils can't tell us what color dinosaurs were. What color do you think they were? Why?

Dinosaurs
Question Cards

Dinosaurs

13 Some reptiles such as snakes, crocodiles, lizards, salamanders, and turtles have hardly changed from their prehistoric ancestors. Why haven't they evolved (changed over time) like many other animals?

Dinosaurs

14 What do you think would happen if someone found a live dinosaur today?

Dinosaurs

15 What special defenses did dinosaurs have to survive?

Dinosaurs

16 The stegosaurus was about the size of an elephant with a brain the size of a walnut. What do you think his typical day was like?

Dinosaurs
Journal Writing Question

Write your response to the question below.
Be ready to share your response.

If dinosaurs did not become extinct, how would the world be different today?

Dinosaurs
Question Starters

Use the question starters below to create complete questions. Send your questions to a partner or to another team to answer.

1. How are dinosaurs _____
_____

2. If you were a T-Rex _____
_____

3. What would you do if _____
_____

4. Which dinosaur _____
_____

5. Why did dinosaurs _____
_____

6. How might _____
_____

7. Will we ever _____

8. How could you find out more _____
_____

Endangered Species

higher-level thinking questions

"I have no riches but my thoughts. Yet these are wealth enough for me."

— Sara Teasdale

Endangered Species
Question Cards

Endangered Species

1 How does the economy of an area endanger its animals?

Endangered Species

2 Evaluate the need for lumber as compared to need for a habitat for the spotted owl.

Endangered Species

3 Imagine yourself as the editor of a new magazine about endangered species. What would you choose for your first cover story and why?

Endangered Species

4 As a government official you must decide how to spend 1,000,000 dollars. How much of that money should be spent on protecting local animal habitats? Why?

Endangered Species
Question Cards

Endangered Species

5 What can people do to protect endangered species in their area, or on another continent?

Endangered Species

6 Why is controlled hunting allowed when so many animals are becoming endangered and extinct?

Endangered Species

7 How does the increase of people on Earth affect the animals?

Endangered Species

8 Should drilling and transporting of oil in the ocean be allowed to continue? Why or why not?

54

Higher-Level Thinking Questions for Life and Earth Sciences
Kagan Publishing • 1 (800) 933-2667 • www.KaganOnline.com

Endangered Species
Question Cards

Endangered Species

9 If the rain forests continue to be destroyed, what are the possible worldwide effects?

Endangered Species

10 What might be invented in the future to help endangered species?

Endangered Species

11 How has the media, (TV, movies) helped educate us about our endangered species?

Endangered Species

12 If you could bring back an extinct species, which species would it be and why?

Higher-Level Thinking Questions for Life and Earth Sciences
Kagan Publishing • 1 (800) 933-2667 • www.KaganOnline.com

Endangered Species
Question Cards

Endangered Species

13 The use of pesticides to kill insects affects the bird population. How does that, in turn, affect the human population?

Endangered Species

14 Choose an endangered species and name as many ways as you can to help preserve it.

Endangered Species

15 Describe how the extinction of one species of animal can affect several others.

Endangered Species

16 You are writing a letter to the newspaper about a high-tech transportation system planned for your area. It will bring jobs and tax dollars to your community and will destroy the habitat of a tiny insect. What will your letter say?

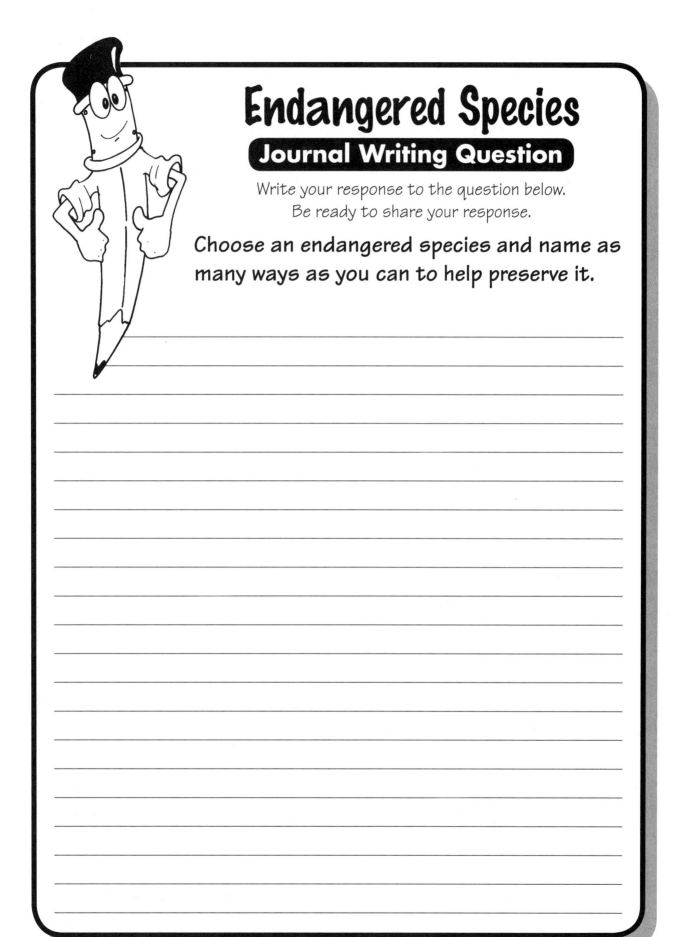

Endangered Species
Journal Writing Question

Write your response to the question below.
Be ready to share your response.

Choose an endangered species and name as many ways as you can to help preserve it.

Endangered Species
Question Starters

Use the question starters below to create complete questions.
Send your questions to a partner or to another team to answer.

1. Why are endangered species

2. What would happen if

3. How would you feel if

4. What can we do

5. How might we

6. Do you think extinction

7. What is more important

8. What are the negative aspects of

Environmental Issues
higher-level thinking questions

"Thought is free.
— William Shakespeare "

Environmental Issues
Question Cards

Environmental Issues

1 When you see a polluted area, how does it make you feel? How could pollution be reduced?

Environmental Issues

2 What are some negative consequences of deforestation? What can be done to limit deforestation?

Environmental Issues

3 To protect furry animals, a new law is proposed. The law makes fur coats illegal. do you support this law? Why or why not?

Environmental Issues

4 You are just elected head of the National Recycling Committee. How do you plan to increase recycling in the country?

Environmental Issues
Question Cards

Environmental Issues

5 What are the causes of acid rain? What are the effects?

Environmental Issues

6 What are three negative effects pollution has on the environment?

Environmental Issues

7 If you were a local politician, not making too much money and big development company offered you millions of dollars to buy land that is supposed to be a wildlife preserve, what would you do?

Environmental Issues

8 The spotted owl is endangered and lives in a forest. If the forest is not cut down, many families will have no work and no money. If the forest is cut down, the owl will become extinct. What should be done?

Higher-Level Thinking Questions for Life and Earth Sciences
Kagan Publishing • 1 (800) 933-2667 • www.KaganOnline.com

Environmental Issues
Question Cards

Environmental Issues

9 If electric cars eliminate emissions and reduce smog pollution, why aren't they more popular yet? What would have to happen to have people switch over to electric cars?

Environmental Issues

10 We know CFCs (chlorofluorocarbons) can damage the ozone layer. CFCs are used in refrigerators and packing materials. Should we stop making refrigerators and packing materials? Why or why not?

Environmental Issues

11 It is feared that we are depleting our ozone and that there will be global warming? What's so bad about warmer weather?

Environmental Issues

12 People who own houses are sometimes against more land development in their area. Should people with homes be allowed to prevent more development in their area or should the land be open to everyone?

Environmental Issues
Question Cards

Environmental Issues

13 If you were the president of the United States, what laws would you propose to protect our environment?

Environmental Issues

14 You live by the beach and your city wants to put up a nearby amusement park. You love amusement parks but know it means they'll have to build a massive sea wall that will eventually cause erosion and loss of beach. Do you support the amusement park or are you against it? Why?

Environmental Issues

15 Who should monitor companies to make sure they are obeying environmental regulations? What should happen if they are found to be in violation?

Environmental Issues

16 What should we do with our toxic waste?

Environmental Issues

Journal Writing Question

Write your response to the question below.
Be ready to share your response.

If you were the president of the United States, what laws would you propose to protect our environment?

Environmental Issues
Question Starters

Use the question starters below to create complete questions. Send your questions to a partner or to another team to answer.

1. What could we do

2. What are the effects of

3. How is pollution

4. Why is the ozone

5. How does recycling

6. What would happen if

7. What could you invent to

8. What do you know about

Higher-Level Thinking Questions for Life and Earth Sciences
Kagan Publishing • 1 (800) 933-2667 • www.KaganOnline.com

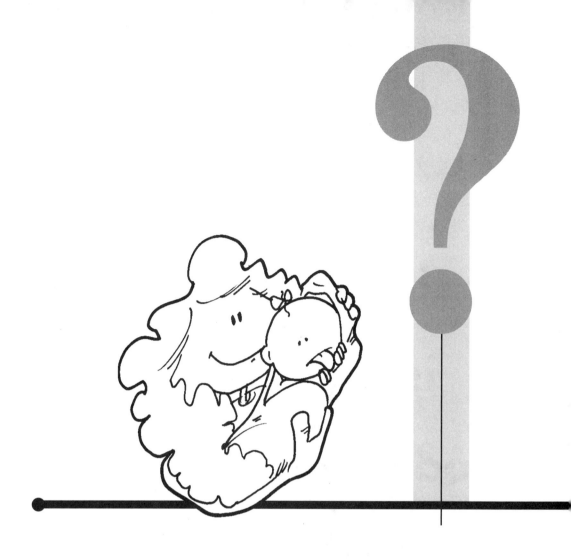

Growing and Aging

higher-level thinking questions

"**Today, the greatest single source of wealth is between your ears. Today, wealth is contained in brainpower, not brute power.**"

— Brian Tracy, Maximum Achievement

Growing and Aging
Question Cards

Growing and Aging

1 Robert Wadlow of the U.S. was the tallest person ever recorded (8 ft. 11.1 in.). What would be three advantages and three disadvantages of being that tall?

Growing and Aging

2 When we are five years old, a year seems like forever. When we get older, they seem to fly by. Why might this happen?

Growing and Aging

3 Women tend to live longer than men. What are some possible explanations?

Growing and Aging

4 As we get old, our hair becomes gray, gets thin or falls out. Our muscles sag and our skin wrinkles. What are some things people do to fight the signs of getting old? Will you?

Growing and Aging
Question Cards

Growing and Aging

5 If you could be any age for the rest of your life, what age would you choose? Why?

Growing and Aging

6 Some people look forward to getting older. Some people fear it. How do you feel?

Growing and Aging

7 As you grow and get older, your body is in a constant state of change. What changes have you noticed?

Growing and Aging

8 At 1,400 pounds, Jon Minnoch was the heaviest human on record. What would it be like to keep growing to be that heavy?

Higher-Level Thinking Questions for Life and Earth Sciences
Kagan Publishing • 1 (800) 933-2667 • www.KaganOnline.com

Growing and Aging
Question Cards

Growing and Aging

9 Is it true that older people are wiser? Why or why not?

Growing and Aging

10 Today, life expectancy is over 70. In the past and it was much, much lower. Why do you think people are living longer today than ever before?

Growing and Aging

11 Myth has it that if you drink from the fountain of youth, you live forever. What would you do if you found the fountain of youth?

Growing and Aging

12 How do babies and old people differ? How are they similar?

Growing and Aging
Question Cards

Growing and Aging

13 If you could change your height, would you? Why or why not?

Growing and Aging

14 Some insects only live for a few days. Some animals only live for a few years. What would it be like to go through your entire life cycle in one year?

Growing and Aging

15 Have you ever been close to anyone who has died? How did you deal with the loss?

Growing and Aging

16 Have you seen pictures or videos of yourself as a baby? What did you think? What is your earliest memory?

Higher-Level Thinking Questions for Life and Earth Sciences
Kagan Publishing • 1 (800) 933-2667 • www.KaganOnline.com

Growing and Aging
Journal Writing Question

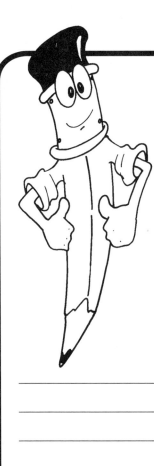

Write your response to the question below.
Be ready to share your response.

Myth has it that if you drink from the fountain of youth, you live forever. What would you do if you found the fountain of youth?

___
___
___
___
___
___
___
___
___
___
___
___
___
___
___
___
___
___

Growing and Aging
Question Starters

Use the question starters below to create complete questions.
Send your questions to a partner or to another team to answer.

1. When you were a baby

2. What general principle

3. If you were

4. Why do people

5. Would you rather

6. How do you feel about

7. What similarities

8. When you get older

Higher-Level Thinking Questions for Life and Earth Sciences
Kagan Publishing • 1 (800) 933-2667 • www.KaganOnline.com

Health and Nutrition
higher-level thinking questions

"The mind is not a vessel to be filled, but a fire to be ignited."

— Plutarch

Health and Nutrition
Question Cards

Health and Nutrition

1 Why is it important to look after your body?

Health and Nutrition

2 What would happen to your teeth if you never brushed or flossed? Describe the sequence of events that might happen.

Health and Nutrition

3 How do you feel if you don't get enough sleep? How many hours of sleep do you need to feel great?

Health and Nutrition

4 Do you have an exercise routine or play a sport? If so, does it keep you fit? If not, what sport or exercise routine would you like to do to keep fit?

Health and Nutrition
Question Cards

Health and Nutrition

5 What would you consider "good hygiene"?

Health and Nutrition

6 Why is smoking so hazardous to your health? Would you smoke even if you knew how bad it was for your health?

Health and Nutrition

7 Your body has powerful defenses against disease and has the ability to repair itself. Give some examples.

Health and Nutrition

8 What would you consider a healthy, balanced dinner? What is a balanced lunch? What is a balanced breakfast?

Higher-Level Thinking Questions for Life and Earth Sciences
Kagan Publishing • 1 (800) 933-2667 • www.KaganOnline.com

Health and Nutrition
Question Cards

Health and Nutrition

9 Do you eat a lot of junk food, or do you try to eat healthy? Why do you think junk food is called "junk" food? How does junk food effect your body?

Health and Nutrition

10 Do you wash your hands before eating? Why is it important?

Health and Nutrition

11 What are some negative consequences of drinking alcohol or taking drugs?

Health and Nutrition

12 What might happen if you left all your food uncovered and unprotected?

Health and Nutrition
Question Cards

Health and Nutrition

13 What is the worst sickness you ever had? How did you feel?

Health and Nutrition

14 How do people get sick? What are some things you can do to minimize your risk of getting sick?

Health and Nutrition

15 What happens to your body if you don't get the exercise you need?

Health and Nutrition

16 What effect do vitamins have on your health?

Higher-Level Thinking Questions for Life and Earth Sciences
Kagan Publishing • 1 (800) 933-2667 • www.KaganOnline.com

Health and Nutrition
Journal Writing Question

Write your response to the question below.
Be ready to share your response.

What would you consider "good hygiene"?

Health and Nutrition
Question Starters

Use the question starters below to create complete questions.
Send your questions to a partner or to another team to answer.

1. Why is it important

2. How is your health

3. What might happen if

4. When you get sick

5. How would you describe

6. How do you decide

7. What foods

8. Is exercise

Human Body
higher-level thinking questions

> "What we want to see is the child in pursuit of knowledge, and not knowledge in pursuit of the child."
>
> — George Bernard Shaw

Human Body Question Cards

Human Body

1 If you could make one improvement on the human body, what would it be? Describe your improvement.

Human Body

2 How would life be different if we walked on all fours?

Human Body

3 How important are your thumbs? How would your life be different if you had no thumbs? What would be different today if humans never had thumbs?

Human Body

4 How are humans and monkeys similar? How are they different? Compare and contrast a human's body to a monkey's body.

Higher-Level Thinking Questions for Life and Earth Sciences
Kagan Publishing • 1 (800) 933-2667 • www.KaganOnline.com

Human Body
Question Cards

Human Body

5 Have you ever seriously injured your body? If so, what did you do? If not, what injury would you fear the most? Why?

Human Body

6 What would your life be like if you had a brain the size of a bird's brain?

Human Body

7 How would the human body be different if we had no bones at all?

Human Body

8 What's more important, your brain or your heart? Why?

Higher-Level Thinking Questions for Life and Earth Sciences
Kagan Publishing • 1 (800) 933-2667 • www.KaganOnline.com

Human Body
Question Cards

Human Body

9 Why do humans have blood?

Human Body

10 Are you afraid of your body aging? What are some things people do to battle the appearance of aging on their bodies?

Human Body

11 If you had to be shorter or taller, which would you choose? Explain your choice.

Human Body

12 Complete the following sentence. "The human body is like_____because..."

Human Body
Question Cards

Human Body

13 You just ate dinner. Trace the path your dinner takes through your digestive system.

Human Body

14 How will the human body be different 10 thousand years from now?

Human Body

15 People often have their appendix, tonsils, and wisdom teeth taken out. If we don't need these parts, why do you think we have them?

Human Body

16 Why do we breathe? Why can't we breathe under water like fish?

Higher-Level Thinking Questions for Life and Earth Sciences
Kagan Publishing • 1 (800) 933-2667 • www.KaganOnline.com

Human Body
Journal Writing Question

Write your response to the question below.
Be ready to share your response.

If you could make one improvement on the human body, what would it be? Describe your improvement.

Human Body
Question Starters

Use the question starters below to create complete questions.
Send your questions to a partner or to another team to answer.

1. How is your body

2. What could you invent

3. What would happen if

4. What would life be like if we had (machine or animal part) instead of (body part)

5. If you had no hands

6. What is the most important

7. How are your bones

8. What do you wonder about

Natural Disasters
higher-level thinking questions

"Learning is by nature curiosity...prying into everything, reluctant to leave anything, material or immaterial unexplained."

— Philo

Natural Disasters
Question Cards

Natural Disasters

1 Have you ever been in a fire, earthquake, tornado, hurricane, or flood? If so, describe what it was like. If not, describe what one might be like.

Natural Disasters

2 What are the causes and effects of an earthquake?

Natural Disasters

3 Which natural disaster scares you most? Which one scares you least? Describe why.

Natural Disasters

4 Why are there natural disasters? Are there any advantages of a natural disaster?

Natural Disasters
Question Cards

Natural Disasters

5 What is the difference between a disaster and a natural disaster?

Natural Disasters

6 How is a hurricane like a tornado? How are they different? Compare and contrast the two.

Natural Disasters

7 What general statements can you make about natural disasters?

Natural Disasters

8 Do you think humans will ever be able to accurately predict natural disasters? Will we ever be able to control them?

Higher-Level Thinking Questions for Life and Earth Sciences
Kagan Publishing • 1 (800) 933-2667 • www.KaganOnline.com

Natural Disasters
Question Cards

Natural Disasters

9 What information would you like to know about natural disasters?

Natural Disasters

10 Which natural disaster do you think has the most potential to harm people? Explain why.

Natural Disasters

11 The highest tsunami ever recorded is thought to have reached 280 ft. What would happen if that wave hit your city?

Natural Disasters

12 Pick one natural disaster. How would you describe it to someone who's never heard of it before?

Natural Disasters
Question Cards

Natural Disasters

13 What are the causes of a flood? What are the effects?

Natural Disasters

14 Why do we call natural events on Earth, natural disasters?

Natural Disasters

15 Some people believe natural disasters are Mother Nature's way of telling us something. Do you agree or disagree?

Natural Disasters

16 If your town was flattened by an earthquake, would it pull people apart or bring them together?

Natural Disasters
Journal Writing Question

Write your response to the question below.
Be ready to share your response.

Have you ever been in a fire, earthquake, tornado, hurricane, or flood? If so, describe what it was like. If not, describe what one might be like.

Natural Disasters
Question Starters

Use the question starters below to create complete questions.
Send your questions to a partner or to another team to answer.

1. How might an earthquake

2. Does a volcano

3. Why are natural disasters

4. If your house caught on fire

5. Do you think a twister

6. What would you do if

7. How is a flood

8. Why do you think

Nature
higher-level thinking questions

> "It is a glorious fever, that desire to know."
>
> —Edward Bulwer Lytton

Nature Question Cards

Nature

1. Picture the most beautiful place in nature you can imagine. Describe it. What makes it so beautiful to you?

Nature

2. How would you feel about camping out in nature for an entire year? What would you do?

Nature

3. If you were a natural scientist, would you study plants or would you study animals? Why?

Nature

4. What do you think is the biggest environmental problem we face today? Why?

Nature Question Cards

Nature

5 Scientists attempt to categorize and classify the things they find in nature. What would your category system for nature be like?

Nature

6 Complete the following sentence. "Nature is…"

Nature

7 What impact does civilization have on nature?

Nature

8 What are some differences between how animals get their nutrition and how humans get their nutrition? What are some similarities?

Nature
Question Cards

Nature

9 What are the advantages of zoos and aquariums? What are the disadvantages?

Nature

10 Do you think there is really someone or something called, "Mother Nature?" If so, describe Mother Nature. If not, why might we use the term?

Nature

11 Would you rather visit the ocean, lake, desert, mountain, rain forest or space. Describe why.

Nature

12 What are the differences between a natural item and a non-natural item?

Nature
Question Cards

Nature

13 How are plants and animals interdependent (depend on each other)?

Nature

14 Every week plants and animals are becoming extinct. What are some things we can do to help protect endangered plants and animals?

Nature

15 What would happen if all the sharks became extinct?

Nature

16 What plant or animal are you most like? Explain why.

Nature

Journal Writing Question

Write your response to the question below.
Be ready to share your response.

Picture the most beautiful place in nature you can imagine. Describe it. What makes it so beautiful to you?

Nature
Question Starters

Use the question starters below to create complete questions.
Send your questions to a partner or to another team to answer.

1. What is the relationship between

2. Do you think nature

3. How could you test

4. If plants and trees

5. If you could eliminate

6. How are animals

7. What might happen if

8. What do you think about

Oceans
higher-level thinking questions

> "One's mind, once stretched by a new idea, never regains its original dimensions."
>
> — Oliver Wendell Holmes

Oceans
Question Cards

Oceans

1 Once the entire world was covered with water. Now it is estimated that oceans make up about three quarters of the earth's surface. Do you think eventually there will be no oceans left?

Oceans

2 How would life be different if you lived underneath the sea?

Oceans

3 Does the ocean scare you or intrigue you? Explain.

Oceans

4 If you could be any oceanic animal, which would you be? Why?

Oceans
Question Cards

Oceans

5 Do you think we should have aquariums and ocean theme parks for public education and entertainment or do you think it's not fair to to keep the animals in tanks? Explain your position.

Oceans

6 How is the ocean different from lakes and rivers? How is it similar?

Oceans

7 Should there be laws restricting the number of fish people should be allowed to take out of the ocean? Why or why not?

Oceans

8 There may be sea creatures lurking in the deep that we haven't discovered yet. What might they look like?

Oceans
Question Cards

Oceans

9 Should the ocean belong to any country?

Oceans

10 How can we reduce ocean pollution? Describe three ideas.

Oceans

11 Have you ever been snorkeling or scuba diving? If so, what did you see? If not, what would you like to see?

Oceans

12 What are some risks people should know about when entering the ocean?

Oceans
Question Cards

Oceans

13 Tide books list the tides for the entire year. How do people know what the tides are going to do in advance? Where does all the water go when the tide is low?

Oceans

14 Why do you think waves are sometimes very big, but sometimes there are no waves at all?

Oceans

15 Have you done any ocean sports? If so, describe your favorite one. If not, describe one you'd like to try.

Oceans

16 Would you rather study the animals of the ocean or the behavior of the ocean itself? Why?

Oceans

Journal Writing Question

Write your response to the question below. Be ready to share your response.

Would you rather study the animals of the ocean or the behavior of the ocean itself? Why?

Oceans
Question Starters

Use the question starters below to create complete questions. Send your questions to a partner or to another team to answer.

1. What would happen if

2. In your life

3. How are waves

4. If you went to an aquarium

5. Which ocean animal

6. How do humans

7. Why might tide pools

8. If all the fish in the ocean

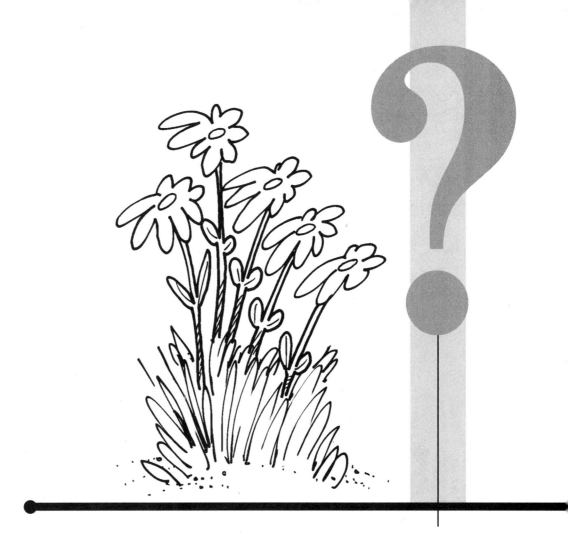

Plants and Trees

higher-level thinking questions

"The whole art of teaching is only the art of awakening the natural curiosity of young minds..."

— Anatole France

Plants and Trees
Question Cards

Plants and Trees

1 How are plants like animals? How are they different? Compare and contrast plants and animals.

Plants and Trees

2 Are you more like a rosebush, an oak tree, a sunflower, or ivy? Describe.

Plants and Trees

3 Fruits and vegetables were put on Earth for humans to eat. Do you agree or disagree with this statement?

Plants and Trees

4 What do you already know about plants? What would you like to know?

Plants and Trees
Question Cards

Plants and Trees

5 What is your favorite flower? What makes it your favorite?

Plants and Trees

6 What are some ways we can reduce the number of trees cut down each year?

Plants and Trees

7 What are some of the many ways humans use plants? Describe five or more.

Plants and Trees

8 How is a seed like an egg? How is it different? Compare and contrast the two.

Plants and Trees
Question Cards

Plants and Trees

9 Imagine you own a palm tree business. You grow your own palm trees and sell them for $40 per foot. If you wanted to grow tall palm trees, what would you do?

Plants and Trees

10 Do plants have feelings? How do you know? If you don't know, what test could you do to find out?

Plants and Trees

11 Why might some plants have spiny leaves, thorny stems or prickly branches? Give some examples.

Plants and Trees

12 How would you describe the relationship between humans and plants?

Plants and Trees
Question Cards

Plants and Trees

13 Every year millions of trees are cut down and used for Christmas trees for a few weeks, then left to die. Do you think that is okay? Is there an alternative?

Plants and Trees

14 "Green thumb" is an expression used for people who are good with plants. Invent a new expression for people who are good with animals.

Plants and Trees

15 Many plants make their own food from water and carbon dioxide using a process called photosynthesis which requires sunlight. How is photosynthesis different than how you get your energy?

Plants and Trees

16 Imagine you opened a nursery and sold plants, trees and flowers. You had enough space to set up five rows. Describe what you would have in each row.

Plants and Trees
Journal Writing Question

Write your response to the question below.
Be ready to share your response.

How are plants like animals? How are they different? Compare and contrast plants and animals.

Plants and Trees
Question Starters

Use the question starters below to create complete questions.
Send your questions to a partner or to another team to answer.

1. What is your favorite

2. How are plants different

3. If you were a tree

4. How do humans

5. If there were no plants

6. How could you symbolize

7. What song

8. If you wanted to test

Rain Forest

higher-level thinking questions

> The art of teaching is the art of assisting discovery.
>
> — Mark Van Doren

Rain Forest
Question Cards

Rain Forest

1 Have you ever been to a tropical rain forest? If so, describe it. What did you like best? If not, would you like to visit one? What would you want to see most?

Rain Forest

2 What would happen if the tropical rain forests of the world were cut down?

Rain Forest

3 Do you think natives to the rain forest have the right to do as they please with the land of the rain forest, or should they have laws? Explain.

Rain Forest

4 What can you do to help save the rain forests?

Rain Forest
Question Cards

Rain Forest

5 Are you more like a toucan, howler monkey, jaguar, or sloth? Explain.

Rain Forest

6 How do you think the daily life of a rain forest native is different from yours? How is it similar?

Rain Forest

7 What are three reasons people would want to clear rain forest land?

Rain Forest

8 List five products that come from the rain forest.

Rain Forest Question Cards

Rain Forest

9 The rain forest is sometimes called the "lungs of the earth." What is another good metaphor for the rain forest?

Rain Forest

10 Rain forest land is often cleared for raising cattle. What is an alternative?

Rain Forest

11 Why do you think most of the world's rain forests are near the equator?

Rain Forest

12 Many rain forest plants may hold medical breakthroughs, but fewer than 1 percent have been studied for medical purposes. What might happen if we studied them all? What will happen if the plants become extinct?

Rain Forest Question Cards

Rain Forest

13 Half of all living things on Earth live in rain forests? Why do you think this is true?

Rain Forest

14 How does the climate of the rain forest compare to where you live?

Rain Forest

15 Many rain forest trees are cut down for hardwood. Hardwood products (such as furniture) should be banned. Do you agree or disagree with this statement? Explain.

Rain Forest

16 Should wealthy countries pay rain forests countries to preserve the rain forest?

Rain Forest

Journal Writing Question

Write your response to the question below. Be ready to share your response.

What would happen if the tropical rain forests of the world were cut down?

Rain Forest
Question Starters

Use the question starters below to create complete questions. Send your questions to a partner or to another team to answer.

1. How are tropical rain forests

2. If you went to a rain forest

3. What is the importance

4. Why are rain forests

5. What products

6. What is an alternative

7. What are the effects of

8. If you were a rain forest animal

Seasons
higher-level thinking questions

"...what is education but a process by which a person begins to learn how to learn."

— Peter Ustinov

Seasons
Question Cards

Seasons

1 What is your favorite season? Why?

Seasons

2 What special things happen only in the spring?

Seasons

3 What would the world be like if the seasons didn't change?

Seasons

4 How is summer different from winter where you live?

Seasons
Question Cards

Seasons

5 What do you know about seasons?

Seasons

6 Seasons are cyclical. That means winter, spring, summer, and autumn always happen in the same sequence. What other things are cyclical? Describe how.

Seasons

7 The word "season" has many meanings. One meaning relates to the time of the year. What is another way you've heard the word "season" used and what does it mean? How is the other meaning different? How is it similar?

Seasons

8 Why doesn't winter follow summer?

134

Higher-Level Thinking Questions for Life and Earth Sciences
Kagan Publishing • 1 (800) 933-2667 • www.KaganOnline.com

Seasons
Question Cards

Seasons

9 How do the seasons affect the way you dress? Describe how you dress in each season.

Seasons

10 Do you think the seasons affect your moods? If so, describe your general mood with each season.

Seasons

11 Most parts of the country have daylight savings. In fall we set our clocks back an hour. In spring, we turn our clocks an hour ahead. Do you prefer it darker later or lighter earlier? Explain.

Seasons

12 The season is determined by Earth's location in relation to the sun. Show someone how Earth's position is different in the winter than in the summer?

Seasons
Question Cards

Seasons

13 When you think of summer, what comes to mind?

Seasons

14 Some things are only "in season" once a year. What things do you know about that have a season? Describe them.

Seasons

15 The same season is called "fall" by some people and "autumn" by others. Which do you prefer. Why?

Seasons

16 Things change with each passing season. Describe three things that change from season to season.

Seasons

Journal Writing Question

Write your response to the question below.
Be ready to share your response.

How is summer different from winter where you live? Think weather, food, clothes, sports, mood, plants…

Seasons
Question Starters

Use the question starters below to create complete questions.
Send your questions to a partner or to another team to answer.

1. What is your favorite

2. What changes

3. If it was always

4. What might happen if

5. Will we ever

6. If there were no seasons

7. What general principle

8. What do you wonder about

Senses
higher-level thinking questions

> **Let our teaching be full of ideas. Hitherto it has been stuffed only with facts.**
>
> — Anatole France

Senses
Question Cards

Senses

1 If you had to lose one sense, which one would you choose? Why?

Senses

2 What would you do if you had the eyesight of an eagle?

Senses

3 In what ways do our senses help us survive?

Senses

4 How would your life be different if you were blind?

Senses
Question Cards

Senses

5 Do you think there's such a thing as a sixth sense or extrasensory perception (ESP)? Describe.

Senses

6 If you could have one sense be super strong, which sense would you choose? Describe how you'd use it.

Senses

7 How do your senses work together to help you make "sense" of the world?

Senses

8 What sense do you rely on most? Explain.

Senses
Question Cards

Senses

9 The word "sense" can also mean logic or wisdom. For example, someone may have a lot of common sense. Or an idea seems senseless. Why do you think the same word has these other meanings?

Senses

10 How important is your sense of taste to you?

Senses

11 How would your life be different if you had no senses?

Senses

12 Describe a situation you've been in that involved all your senses at once.

Higher-Level Thinking Questions for Life and Earth Sciences
Kagan Publishing • 1 (800) 933-2667 • www.KaganOnline.com

Senses
Question Cards

Senses

13 If you had no sense of touch, it could be considered an advantage because you wouldn't feel pain. List three disadvantages.

Senses

14 Beethoven the famous composer became deaf, yet still created beautiful music. His last words were, "I shall hear in heaven." In what way is hearing an important sense to you?

Senses

15 Sharks have a terrific sense of smell and owls can see at night. Why do you think some animals have really strong senses?

Senses

16 Would you like it if you could hear people's thoughts? If so, whose thoughts would you listen to? If not, why not?

Senses

Journal Writing Question

Write your response to the question below.
Be ready to share your response.

If you could have one sense be super strong, which sense would you choose? Describe how you'd use it.

Senses
Question Starters

Use the question starters below to create complete questions.
Send your questions to a partner or to another team to answer.

1. If you had no senses

2. Why is hearing

3. What if your sense of smell

4. How does taste

5. How do you use

6. Do you think sight

7. Why are your senses

8. Is touch

Weather and Climate

higher-level thinking questions

"If he is indeed wise [the teacher] does not bid you enter the house of his wisdom, but rather leads you to the threshold of your own mind."

— Kahlil Gibran, The Prophet

Weather and Climate
Question Cards

Weather and Climate

1 What is your favorite kind of weather? Would you like it if it was always your favorite weather outside?

Weather and Climate

2 What might happen if it didn't stop raining? What might happen if it never rained again?

Weather and Climate

3 What would you have to do if it regularly hailed the size of basketballs?

Weather and Climate

4 You've invented a machine that gives perfect weather predictions every day. What will you do with it?

Higher-Level Thinking Questions for Life and Earth Sciences
Kagan Publishing • 1 (800) 933-2667 • www.KaganOnline.com

Weather and Climate
Question Cards

Weather and Climate

5 As technologically advanced as we are today, meteorologists still have a hard time predicting the weather. Why do you think weather is so hard to predict? Do you think they will ever be perfect? Why or why not?

Weather and Climate

6 How would you describe the climate of where you live? If you could change one thing about the climate in which you live, what would it be?

Weather and Climate

7 If you had a weather machine that allowed you to control the weather, what would you do with it?

Weather and Climate

8 Describe three disadvantages of living in the snow. Describe three advantages.

Higher-Level Thinking Questions for Life and Earth Sciences
Kagan Publishing • 1 (800) 933-2667 • www.KaganOnline.com

Weather and Climate
Question Cards

Weather and Climate

9 Climate is divided into zones such as: Mountain, Tropical Wet and Dry, Desert, Subtropical Dry Summer, Continental Moist, Oceanic Moist, Subarctic, Polar. Why does climate vary so much from place to place?

Weather and Climate

10 What is your least favorite weather? Why?

Weather and Climate

11 Compare and contrast a rainy day and a snowy day.

Weather and Climate

12 Have you ever been in really scary weather? If so, describe it. If not, what weather would be most scary?

Weather and Climate
Question Cards

Weather and Climate

13 Would you rather be a meteorologist who studies the weather or a TV personality who broadcasts the weather? Why?

Weather and Climate

14 What is the weather like today? How do you feel today? Does the weather affect your mood? If so, how so?

Weather and Climate

15 What makes the weather change from day to day? Would you prefer the weather to be the same every day?

Weather and Climate

16 What does, "It's raining cats and dogs," mean? Make up a new expression to describe another weather pattern.

Weather and Climate
Journal Writing Question

Write your response to the question below.
Be ready to share your response.

If you had a weather machine that allowed you to control the weather, what would you do with it?

Weather and Climate
Question Starters

Use the question starters below to create complete questions. Send your questions to a partner or to another team to answer.

1. How is the weather

2. What invention

3. How could you test to

4. What is the relationship between

5. How can you explain

6. Would you prefer

7. What plans would you make if

8. How could you best represent